$\mathscr{S}$UNS

—⁓— AND —⁓—

$\mathscr{M}$OONS

$\mathscr{S}$UNS

— AND —

$\mathscr{M}$OONS

NEW
HOLLAND

First published in the UK in 1995 by
New Holland (Publishers) Ltd
Chapel House, 24 Nutford Place
London W1H 6DQ

ISBN 1 85368 534 8

Editorial Direction: Yvonne McFarlane
Design: Peter Crump
Special photography: Shona Wood
Text researcher: Joanna Ryde
Craft designs and additional research: Chris Fox

Reproduction by Dot Gradations Ltd,
South Woodham Ferrers,
Chelmsford, Essex
Printed and bound in Singapore by
Tien Wah Press (Pte) Ltd

Contents

Heavenly Zodiacs

The Sun came up upon the left
Out of the sea came he!
And he shone bright, and on the right
Went down into the sea.

Samuel Taylor Coleridge (1772-1834),
The Rime of the Ancient Mariner (1798)

$\mathcal{A}$nd God set them in the firmament of the heaven
to give light upon the earth.

Genesis, I:17

$\mathcal{S}$o sicken waning moon too near the sun,
And blunt their crescents on the edge of day.

John Dryden (1631-1700), *Annus Mirabilis*

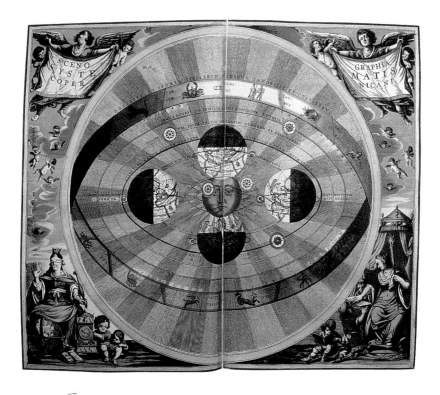

$\mathcal{I}$n my studies of astronomy and philosophy I hold this opinion
about the universe, that the Sun remains fixed in the centre of the
circle of heavenly bodies, without changing its place; and the
Earth, turning upon itself, moves around the Sun.

Galileo Galilei (1564-1642), *Letter to Cristina di Lorena* (1615)

Glittering Moons Decorations

*F*estive decorations have been in vogue since the turn of the century when trees were decked with home-made trimmings of shimmering paper, charms, fruit and bonbons. These glittering moons and stars can be hung on the tree or in your window for a traditional welcome.

*M*ake cardboard templates of a crescent moon and two differently sized star shapes. Draw round them to make an outline on stiff white card and cut out these pieces with a craft knife. Using a small artist's brush, cover both surfaces and sides of the shapes with gold acrylic paint and allow them to dry.

*U*sing a needle or the point of a sharp knife, make small holes in the edges of the moon and in one of the points of each of the stars. Push lengths of gold metallic thread through the holes to join the images together. Knot the yarn at the back of each hole to secure them. Thread an extra length of fine cord through the top of the moon for hanging.

*L*ine a large box with kitchen foil and pour in tiny gold star decorations. Coat one side of each moon and star shape with clear adhesive. Gently place the shapes in the box and, if necessary, shake the box around so the stars stick to them; turn the shapes over and repeat on the other side so that all the decorations are completely covered with stars.

*F*or age and want, save while you may!
No morning sun lasts a whole day.

Benjamin Franklin (1706-90), *Poor Richard's Almanac* (1758)

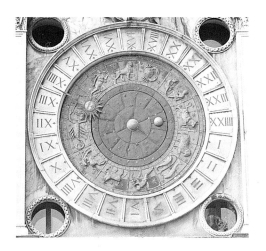

$\mathcal{N}$ow hast thou but one bare hour to live,
And then thou must be damned perpetually!
Stand still, you ever moving spheres of Heaven
That time may cease, and midnight never come.

Christopher Marlowe (1564-93),
The Tragical History of Doctor Faustus (1604)

From the golden limits of the east,
By the ocean richly girdled with pearls,
To the darkened edges of the west,
The fringes of your gleaming garments
Shine in splendour
August and majestic –
And you bathe all the world
In your pure fires.

José de Espronceda, *The Sun*

XVIIII

LE · SOLEIL

XVIII

LA · LUNE

*T*ime will not be ours forever,
He at length our good will sever.
Spend not then his gifts in vain;
Suns that set may rise again;
But if once we lose this light,
'Tis with us perpetual night.

Ben Jonson (1573-1637), Song to Celia

*T*he bodies that occupy the
    celestial vault
These give rise to wise men's
    uncertainties;
Take care not to lose your grip on
    the thread of wisdom,
Since the Powers That Be
    themselves are in a spin.

Ruba'iyat of Omar Khayám

Golden Suns, Silver Moons

The moon was created so that the sun
be not regarded a god.

Jewish Proverb

$\mathcal{N}$othing that is can pause or stay;
The moon will wax, the moon will wane,
The mist and cloud will turn to rain,
The rain to mist and cloud again,
Tomorrow be today.

Henry Wadsworth Longfellow (1807-82), *Keramos* (1878)

Crackle-glaze
Sun Box

*D*esigns made from cut-out paper scraps which are pasted down and varnished are known as decoupage, and this style of decoration can turn a plain object into a Victorian-style work of art. These suns and moons were cut out from wrapping paper. To make your own designs, simply choose images of a suitable size and reduce or enlarge them on a colour copier.

*T*o ensure that your pictures stand out, paint a plain box with acrylic in a lighter shade than the images used for decoration. Cut out the images, trimming off ragged edges. Try out different arrangements on the box before you finally stick them down.

*A*llow the adhesive to dry thoroughly, then apply two coats of clear varnish to seal and protect the design. Add a crackle glaze for an "antique" finish.

*T*ime has left off its coat of wind, cold and rain;
And is decked out
In embroidery of gleaming sun,
Clear and beautiful.

Charles d'Orléans, *Rondel*

*B*ehold the heavens! A devouring flame
Has taken hold of them
The sun with fiery kiss has touched
And set afire their hem.

J. Cahan, *Shekiat Hama*

It rises – passes – on our South,
Inscribes a simple Noon,
Cajoles a moment with the spires,
And infinite, is gone.

Emily Dickinson (1830-86)

*B*usy old fool, unruly Sun
Why dost thou thus,
Through windows, and
Through curtains
Call on us?

John Donne (1571-1631), *The Sun Rising* (c. 1605)

Suns and Moons Album

To make the back of the album, cut a piece of thick cardboard 1/4 in (6 mm) larger all round than the pages. Cut a front board 3/8 in (1 cm) narrower in width than the back board, then cut a strip off the short edge approximately 1 in (2.5 cm) wide. This will form a hinge on the left side, allowing the album to open easily. To make the padding, cut a piece of 1/4 in (6 mm) thick foam to fit the front board exactly (excluding the hinge) and stick it to the top side of the board.

Cut fabric to cover the back board, allowing 3/4 in (2 cm) extra all round to turn in. Repeat for the front board, adding an extra 2 in (5 cm) to the width to return under the hinge and onto the underside of the front board.

Iron paper-backed bonding web to the wrong side of both pieces of fabric, Peel off the paper, lay the fabric face down and position each board in the centre. Make sure the front board is foam-side down and the board and hinge are 3/8 in (1 cm) apart, so that their overall width is the same as the back. Trim the corners off the fabric diagonally and turn in all round using a hot iron to bond the fabric to the boards. Stick cartridge paper to the inside back board to neaten what will be the inside back cover. Repeat for the front board.

Align the left edges of the front and back boards with the cartridge paper and punch two holes through in the centre of the hinge. Slot the ribbon through the holes and tie with a bow.

And I love to see the sun rise blood-crimson.
And watch his spears through the dark clash
And it fills all my heart with rejoicing
And pries wide my mouth with his music
When I see him so scorn and defy peace,
His lone might 'gainst all darkness opposing.

Ezra Pound (1885-1972), *Sestina: Altaforte*

The Art of
Light

There is something haunting in the light of the moon;
it has all the dispassionateness of a disembodied soul,
and something of its inconceivable mystery.

Joseph Conrad (1857-1924), *Lord Jim* (1900)

*H*is last great love was for the sun, which he glorified
in his pictures. Men would have nothing to do with these
pictures and laughed at them. And even the sun did not love
him; it robbed him of his reason and killed him.

Wilhelm Uhde, *Vincent van Gogh* (1947)

*T*he brilliance of the sun does not
remain hidden; at its sight the heart
of the lotus has bloomed.

Malik Muhammad Jayasi, *Padumavati*

*T*he night has a thousand eyes,
And the day but one;
Yet the light of the bright world dies
With the dying sun.

Francis William Bourdillon (1852-1921), *Light*

Heavenly Greetings Cards

Potato cuts are a simple form of block printing using designs cut into the flat surface of a potato sliced in half.

These cards were made using rough watercolour paper. You can make a deckle edge by simply tearing the paper along a ruler.

Cut a large potato in half and dry the cut edges with paper towels. Cut out a moon shape on the flat surface of one of them, cutting into its flesh to a depth of about 1/4 in (6 mm). Apply gold acrylic paint to the design and press the potato onto the paper to stamp out the image.

Repeat the process to form a pattern on the paper. Punch holes in the side of your card and thread with gold ribbon. Tie with a bold bow for an elegant finishing touch.

*I*s it so small a thing
To have enjoy'd the sun,
To have lived light in the spring,
To have loved, to have thought, to have done?

Matthew Arnold (1822-88), *Empedocles on Etna*

*T*he sun extends its splendour over mountains and valleys, waking birds and animals and people. Some fly through the clear air, others through the green valleys or over the high summits, safely and peacefully grazing; some, now that the sun is up, go to their usual work to which nature or necessity has accustomed them.

Garciloso de la Vega, *Eclogues*

*T*urner had strength to bear that tempering
That shatters weaker hearts and breaks their hope
He still pursued his journey step by step
First modestly attired in quiet grey
As well became sincere humility;
Then with a plume of colour he adorned
His simple raiment and so walked awhile
Until at last, like his beloved Sun
He set in forms of strangest phantasy
Coloured with gold and scarlet and the lands
Of his conception grew as dim and vague
As shadows. So his mighty brain declined.

Philip Gilbert Hamerton, *Turner*

ℒight upon Earth

ℒook! The moon is climbing through these shadows.
Beautiful queen of the night,
Your gaze still trembles, but you are
Already disengaging yourself from the dim horizon
And showing yourself clearly.

Alfred de Musset (1810-57), *Memory*

*T*he moon is like a sleeping child
Whose golden locks have fallen
Over his beloved face.

Georg Buchner, *Leonce and Lena*

*M*y face is pale
And full and fair;
And round it,
Beauty spots there are.
By day indeed,
I seem less bright,
And only seen,
Sometimes at night
And when the Sun
Is gone to bed
I then begin
To show my head.

The Moon – *A Whetstone for Dull Wits*
(18th century riddle)

Crescent
Moons
Stencil

*T*ape a piece of stencil acetate over the design and with a waterproof pen, trace the image on to it. Cut away the image using a sharp craft knife. Make sure the acetate is a couple of inches (5 cm) larger all round than the stencil design so you ensure the edges will not split.

To prevent the stencil from moving while you paint, use low-tack masking tape to fix it to the surface you are decorating.

Dip a dry sponge into the paint and remove any excess on a clean piece of paper. To create a delicate, dappled effect, use very little paint and vary the pressure with the sponge. Leave to dry for a few seconds, then peel off the masking tape and move the stencil along, lining it up evenly. When completely dry, the stencil can be varnished for added protection.

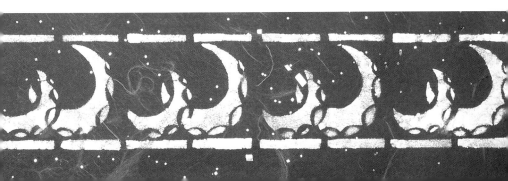

And God made two great lights;
the greater light to rule the day,
and the lesser light to rule the night...

Genesis, I , 16

$\mathcal{T}$here is one glory of the sun,
and another glory of the moon,
and another glory of the stars:
for one star differeth from
another star in glory.

Corinthians I, 15

$\mathcal{A}$t midnight, the moon shines in
the sky like a silver coin on an
azure banner spangled with
golden bees!

Aloysius Bretrand, (1807-41), *Scarbo*

*L*ook for me by moonlight;
Watch for me by moonlight;
I'll come to thee by moonlight,
Though hell should bar the way!

Alfred Noyes (1880-1958), *Forty Singing Seamen*

*H*ow sweet the moonlight sleeps upon this bank!
Here we will sit and let the sounds of music
Creep in our ears: soft stillness and the night
Become the touches of sweet harmony.

William Shakespeare (1564-1616), *Merchant of Venice*

*T*he moon combed its tresses with
a broad-toothed comb of ebony,
releasing a silver shower of
glow-worms on the hills, the meadows
and the woods.

Aloysius Bertrand, (1807-41), *Gaspard of the Night*

The Celestial
Vault

Had I the heavens' embroidered cloths
Enwrought with golden and silver light...

W. B. Yeats (1865-1939)
He wishes for the Cloths of Heaven.

Par sacrifice et prière De nostre lieu Jesus Christ

*F*ull many a morning have I seen
Flatter the mountain tops with sovereign eye,
Kissing with golden face the meadows green
Gilding pale streams with heavenly alchemy.

William Shakespeare (1564-1616), *Sonnets*, 33

*A*nd thou, O Sun! Thou seest all things
And hearest all things
In the daily course.

Homer (c. 900 BC), *The Iliad*, III, 27

Man-in-the-Moon Plaque

The man-in-the-moon plaque was made from regular salt dough rolled out 1/4 in (6 mm) thick. Using a small cake tin as a guide, cut a dough circle. Cut out a cresent shape of similar size from the remaining dough and moisten the back of it with warm water. Press the crescent firmly down on one edge of the dough circle, lining up the outer edges to fit neatly together.

Cut lips, eyelid, nostril and star shapes from the remaining dough, moisten each shape and position them on the crescent. Insert a paperclip in the back of the plaque to make a small loop.

Dry the dough for about 10 hours or overnight in a *very slow* oven 225 °F (110 °C/gas mark 1/4).

When dry, decorate with acrylic paint. For an aged effect after colouring, apply a wash made from burnt umber acrylic thinned with water. Allow to dry and then paint twice back and front with polyurethane varnish.

*T*he sun shall not smite you by day,
Nor the moon by night.

Psalm 21

I would not be the Moon, the sickly thing
To summon owls and bats upon her wing
For when the noble sun is gone away
She turns his nights into a pallid day.
She hath no air, no radiance of her own
That world unmusical of earth and stone
She wakes her dim, uncoloured, voiceless hosts,
Ghost of the sun, herself the sun of ghosts.
The mortal eyes that gaze too long on her
Of Reason's piercing ray defrauded are.
Light in itself doth feed the living brain;
That light, reflected, but makes darkness plain.

Mary Coleridge (1861-1907), *In Dispraise of the Moon* (1908)

*Y*on rising Moon that looks for us again –
How oft hereafter will she wax and wane;
How oft hereafter rising look for us
Through this same garden – and for *one* in vain!

Edward Fitzgerald (1809-83)

$\mathscr{S}$weet the coming on
Of grateful evening mild, then silent night
With this her solemn bird and this fair moon,
And these gems of Heav'n, her starry train.

John Milton (1608-74) *Paradise Lost*

*E*veryone is a moon and has a dark side
which he never shows to anybody.

Mark Twain (1835-1910),
Notebook (1935)

*P*eacefully the moon pauses over the rooftops,
And hovers by the orchards,
Lighting each distant mountain
In a picture of calm.

Giacomo Leopardi

Warm summer sun, shine kindly here;
Warm southern wind, blow softly here;
Green sod above, lie light, lie light –
Goodnight, dear heart, goodnight.
Goodnight.

Mark Twain (1835-1910), *Epitaph for his Daughter*

He who doubts from what he sees
Will ne'er believe, do what you please.
If the Sun and Moon should doubt,
They'd immediately go out.

William Blake (11757-1827), *Book of Thel*

Acknowledgements

The Publishers would like to thank the Bridgeman Art Library (BAL), the E.T.Archive (ETA), AKG London (AKG), Visual Arts Library (VAL), Ancient Art & Architecture (AA&A), Fine Art Photographic Library (FAPL), for the loan of the pictures included in this book.

Front cover, John Atkinson Grimshaw, *Spirit of the Night*; (FAPL); frontispiece, *The Creation of the Sun and Moon*, St. Florentin, nr. Auxerre (BAL); 6, *Splendor Solis* manuscript, Augsburg, 16th Century *Arma Artis*, SMPK, Kupferstichkabinett, Berlin (AKG); 7, Biblioteca Estense Modena; 8, Camille Flammarion, woodcut, *L'atmosphère metréologie populaire*, Paris 1888 (AKG); 9, Scenographia: *Systematis Copernicani* astrological chart, 1543, British Library, London (BAL); 12, moon clock, courtesy Roger Lascelles Clocks; 13, astrological clock: Venice Clock Tower (AA&A); 14, Sun box, courtesy of On Stage; 15, detail from *Genealogy*, 18th century Cuzco School painting of Manco Capac Inca, Pedro de Osma Museum , Lima (ETA); 16, tarot cards from *Le Tarot de Marseilles*; 17, Antonio Cicognara, *The Moon*, tarot cards, 16th century, Carrara Academy Bergamo (ETA); 18, detail from Hall of Chivalry, Kea, Lacko Castle , Sweden (ETA); 19, detail from 18th century Cuzco School painting of *Corpus Christi* procession, Archbishop Palace Museum, Cuzco (ETA); 20 (top), teddy-on-moon tree decoration, courtesy of Chris Fox; (below), Art Deco powder compact, courtesy of The Cartier Collection (below); 21, moon brooch, courtesy of Butler & Wilson; 24, velvet embroidered cushion, courtesy of William Price, photographed by Marie-Louise Avery; 25, glass scent bottle, courtesy of Liberty; 26, detail of entrance gate to Sans Souci, Potsdam, Germany (ETA); 27, (left), *The Rising Sun*, enamelled copper vase, courtesy of Victor Arwas, London; (right), cigarette lighter, courtesy of The Cartier Collection; 30, Art Deco ware, courtesy of John & Carol Ann Phipps; 31, gold sun plaque, courtesy of Fenwick of Bond Street; 32, detail of sun brooch, courtesy of Zsuzsi Morrison; 33, Henri Rousseau, *The Snake Charmer*, Lauro-Giraudon/Musée National d'Art Moderne, Paris (BAL); 34, Vincent van Gogh, *The Sower*, 1888, E. G. Buehrle Collection, Zurich (AKG); 35,

Samuel Palmer, *Harvest Moon* (BAL); 36, picture frame, courtesy of Fenwick of Bond Street, inset,*Akhenaten and Nefertiti Adoring the Solar Disc*, 18th dynasty (ETA); 37 and 74, Solid wood *Sun and Moon*, 19th century German folk art, from the collection of the Kunst & Geschichte, Berlin (AKG); 40, pewter plaque, courtesy of Liberty; 41, pierced tin sun, courtesy of Paperchase; 42, *Sunrise*, from R. Caldecott's *Collection of Pictures and Songs*, Mary Evans Picture Library; 43, Heinrich Oswalt, *Spring*, coloured by Eugen Klimsch (AKG); 44, William Turner, *Sunset*, Tate Gallery, London (VAL); 45, Caspar David Friedrich, *Two Men Contemplating the Moon*, 1819, Castle Pillinitz, Dresden (AKG); 46, majolica plate, Siena 1733, (AA&A); 47, Sir Joseph Noel Paton, *A Dream of Latmos* (FAPL); 49, Edward Robert Hughes, R.A., *Radiant Moon* (FAPL); 52, J. Corbechon, *Marriage of Adam & Eve*, Fitzwilliam Museum, Cambridge (VAL); 53, Johann von Gmünd, woodcut from a calendar, *The Month of May* (AKG); 54-55, Cosmic Garland, courtesy of Paperchase; 56, V. F. Pollet, *Endymion and Selene*, Victoria & Albert Museum, London (VAL); 57, John Simmons, *Titania* (FAPL); 58, 19th century south German folk art, *Mrs Sun*, in wood, from the collection of the Kunst & Geschichte; 59, sun plaque, by Kathy Fillion Ritchie/Providence Cabinet Makers, courtesy of Yvonne McFarlane; 60,*The Creation of the Sun, Moon and Stars*, St. Madeleine, Troyes, France (BAL); 61, *Akenaton in the form of a sphinx* (AA&A); 64 Hrastovije, former Yugoslavia.*Sun and Moon Fresco*, G. Tortoll/(AA&A); 66-67, Turkish *Treatise on Astrology*,: *Cancer*, Bibliothèque Nationale, Paris (ETA); 68, Tintoretto, orig. Jacapo Robusti, *Veritas*, Doge's Palace, Sala del Senato, Venice (AKG); 69 and back cover, *Splendor Solis*, manuscript, Augsburg, 16th century *Conflict Between Sun and Moon*, SMPK, Kupferstichkabinett, Berlin (AKG); 71, star-shaped box, courtesy of Fenwick of Bond Street.

The Publishers have made every effort to identify copyright holders of material included in this book and apologise for any inadvertent omissions, which they will rectify in the event of a reprint.